The Future is Now: How SpaceX's Starship Rocket is Changing the Game.

SCOTT N. MOORE

COPYRIGHT PAGE

TABLE OF CONTENTS

Introduction: The Vision for Starship and Why It Matters

SpaceX's Starship rocket provides a daring and ambitious vision for the future of human space travel. The objective of the Starship program is to construct a completely reusable spaceship that can take both persons and cargo to locations across the solar system, including the Moon, Mars, and beyond.

At its heart, the idea for Starship is about opening up new frontiers in space and increasing humanity's presence beyond Earth. It marks a tremendous step forward in our capacity to explore and populate other planets, as well as a new age of space travel that might alter the way we live and operate in space.

Starship is also a vital component of SpaceX's bigger ambition for making life multi-planetary. By constructing a spaceship that is totally reusable and can transport vast quantities of cargo and personnel, SpaceX intends to make space travel more accessible and inexpensive, which might pave the way for a future when humans are able to create sustainable communities on other planets.

Beyond its possible influence on space travel and colonization, Starship also has huge ramifications for our planet's future. By allowing humans to reach space more quickly and responsibly, Starship might play a crucial role in creating new technologies and businesses that could help solve some of the largest concerns facing our world, from climate change to resource depletion.

Overall, the idea for Starship is one of unlimited possibilities and promise. It symbolizes a new frontier in human exploration and a vital step forward in our ambitions to become a spacefaring civilisation.

The History and Evolution of SpaceX

SpaceX, short for Space Exploration Technologies Corp., was formed in 2002 by entrepreneur Elon Musk with the purpose of revolutionizing space travel and making it more inexpensive and accessible.

The company's early years were marred by a succession of setbacks and disasters, notably the first three flights of its Falcon 1 rocket, which all ended in failure. However, SpaceX persisted, and in 2008, the fourth launch of the Falcon 1 was successful, making SpaceX the first privately financed firm to send a spacecraft into orbit.

In the years that followed, SpaceX continued to make breakthroughs in spaceflight, producing the Falcon 9 rocket, which is capable of carrying cargo and passengers to the International Space Station (ISS). In 2012, SpaceX became the first privately financed corporation to fly a spaceship to the ISS, clearing the door for a new age of commercial spaceflight.

In addition to its work with the Falcon rockets, SpaceX has also made considerable breakthroughs in constructing reusable spacecraft.

In 2015, the business successfully landed the first stage of a Falcon 9 rocket on a drone ship in the Atlantic Ocean, marking a key milestone in the company's ambitions to cut the cost of spaceflight.

More recently, SpaceX has moved its attention to the construction of the Starship rocket, which marks the next phase in the company's attempts to make space travel more affordable and sustainable. Starship is a completely reusable spaceship that is meant to take both persons and cargo to locations across the solar system, including the Moon, Mars, and beyond.

Throughout its inception, SpaceX has been driven by a goal of revolutionizing space travel and exploration, making it more accessible and cheap for people throughout the globe.

With each new success and milestone, the business has continued to push the frontiers of what is possible in spaceflight, setting the path for a future where space is more accessible and open to exploration than ever before.

The Design and Development of Starship

The design and development of SpaceX's Starship rocket has been a complex and iterative process, with the company experimenting with a wide range of materials, technologies, and engineering approaches to create a fully reusable spacecraft that can carry both cargo and humans to destinations throughout the solar system.

The current concept of Starship consists of a 50-meter-tall stainless steel spaceship that is propelled by six Raptor engines. The spaceship is planned to transport up to 100 passengers, along with a variety of cargo and research equipment, to destinations including the Moon and Mars.

One of the fundamental elements of Starship's design is its use of stainless steel as a principal construction material. Stainless steel is strong, heat-resistant, and simple to work with, making it an ideal option for a spaceship that has to survive the intense temperatures and pressures of space flight.

The development of Starship has also entailed the introduction of new technologies and engineering methodologies to allow the spaceship to be totally reusable. This includes the use of retractable landing legs and a heat shield that can withstand the extreme heat of reentry.

In addition, the development of Starship has involved extensive testing and experimentation, including a series of test flights that have been conducted at SpaceX's Boca Chica launch facility in Texas. These tests have included both suborbital hops and high-altitude test flights, with the goal of validating the spacecraft's design and performance in real-world conditions.

Throughout the design and development process, SpaceX has remained committed to its vision of creating a fully reusable spacecraft that can transform the economics of space travel and open up new frontiers in space exploration. With each new milestone and achievement, the company has continued to refine its design and push the boundaries of what is possible in spaceflight.

Why the Starship Exploded:

SpaceX's Starship rocket exploded on Thursday, minutes after lifting off from a launchpad in South Texas. The spacecraft, the most powerful ever to launch, failed to reach orbit, but it was not a total failure for the private spaceflight company.

SpaceX's Starship rocket cleared its launch platform but failed to separate from its booster, exploding four minutes after liftoff during an inaugural test flight on Thursday.

Before the launch, Elon Musk, the company's founder, had tamped down expectations, saying it might take several tries before Starship succeeds at this test flight, which was to reach speeds fast enough to enter orbit before splashing down in the Pacific Ocean near Hawaii.

But the launch achieved a number of important milestones, with the rocket flying for four minutes and getting well clear of the launchpad before it started to tumble, culminating in a high-altitude blast. The short flight yielded reams of data for engineers

Despite the setback, SpaceX remains the leading corporation in global spaceflight. Its Falcon 9 rockets have already gone to orbit 25 times in 2023, with the most recent launch finishing safely on Wednesday.

The countdown on Thursday at the launch site in South Texas, near the city of Brownsville, progressed smoothly through the morning until the final half a minute, when it was delayed for a few minutes as SpaceX engineers rectified technical concerns. Employees at SpaceX headquarters in California began applauding loudly as the countdown restarted.

At 9:33 a.m. Eastern time, the 33 engines on the Super Heavy booster erupted in a massive cloud of fire, smoke and dust, and the Starship climbed gently upward. About a minute later, the rocket went through a phase of maximum aerodynamic pressure, one of the essential periods for the launch of any rocket.

The aim for the mission had been to send the spacecraft on one near-complete rotation of the Earth, concluding with a splashdown in the Pacific, a few of hundred kilometres north of Hawaii.

There was no anticipation that the ship or Super Heavy would be retrieved. But long term, this is the plan. The objective is to land both halves, refuel them and launch again - time and time again.

If this can be realized, it will be transformational.

Starship has a projected payload performance to orbit of more than 100 tonnes a mission. When this is connected to the cheap cost of operation - essentially, simply the cost of gasoline - it should open the door to an exciting future.

The Technical Challenges of Building Starship: Materials, Engines, and Propulsion Systems

Building SpaceX's Starship has brought a number of substantial technological obstacles, notably in the fields of materials, engines, and propulsion systems.

One of the primary issues in creating Starship has been the selection of materials for the spaceship. Stainless steel, which has been employed as the major construction material for the rocket, is quite hefty, which may make it more difficult to reach the requisite levels of performance and efficiency needed for interplanetary travel.

However, SpaceX has addressed this difficulty with a number of design advances, including the use of a relatively thin steel skin for the spaceship, which helps to lower its total weight.

Another important technological hurdle has been the creation of the Raptor engines that power Starship. These engines are intended to burn a blend of methane and oxygen, which gives a variety of benefits over typical rocket fuels.

However, the creation of these engines has required tremendous expenditure in research and development, as well as the development of new manufacturing processes to create the intricate components necessary for the engines.

In addition to the engines, the propulsion technology employed by Starship also faces a variety of technological obstacles. The spacecraft is planned to employ a mix of standard rocket engines and a series of smaller thrusters, which give extra control and mobility during flight.

This needs precise coordination between the engines and thrusters, as well as complex software systems to guarantee that the spacecraft is able to accomplish its target trajectory.

Overall, the technological problems involved in developing Starship have been enormous, requiring significant expenditure in research, development, and testing.

However, SpaceX's drive to invention and experimentation has helped the business to overcome many of these hurdles and make major progress in the creation of a completely reusable spaceship that has the potential to revolutionize space travel and exploration.

The Future of Space Travel: How Starship Will Change Space Exploration

SpaceX's Starship is set to be a game-changer in the field of space exploration, having the potential to substantially alter the future of space travel in a variety of ways.

One of the primary ways that Starship will transform space exploration is by making it more accessible and economical. The totally reusable aspect of the spacecraft, along with its high level of performance and capacity, means that it has the potential to dramatically cut the cost of space travel, making it more accessible to a larger variety of organizations and people.

In addition, the capabilities of Starship will allow a spectrum of new missions and prospects for space exploration. For example, the spacecraft's enormous capacity and lengthy endurance will make it perfect for carrying out protracted missions to the Moon, Mars, and beyond. It might also be used to launch scientific missions, such as telescopes or other devices, into outer space.

Another important influence of Starship on space flight will be its capacity to stimulate innovation and cooperation in the sector. The creation of the spacecraft has already fostered a variety of new technical improvements and techniques, and its deployment by a broad range of organizations might promote more innovation and cooperation in the sector.

Overall, the future of space travel with Starship is fascinating and full of possibility. As the spacecraft continues to be built and tested, it is possible that we will see many more imaginative uses and applications for this pioneering technology.

The Environmental Impact of Starship and the Future of Sustainable Space Travel

The development of Starship and the rising desire in space travel and exploration raise issues about the environmental effect of spaceflight and the need for sustainable practices in the sector.

The construction of rockets and spacecraft, as well as the emissions from rocket engines, have the potential to contribute to climate change and other environmental problems. However, the adoption of innovative technology and sustainable practices in the building and operation of spacecraft like Starship might assist to offset these consequences.

One significant method that Starship and other spacecraft might lessen their environmental effect is via the utilization of renewable energy sources. For example, solar panels might be utilized to power onboard equipment and lessen dependency on conventional fuel sources. Additionally, the development of more efficient engines and propulsion systems might assist to cut emissions and enhance overall efficiency.

Another major aspect for sustainable space flight is the proper handling of garbage and other environmental consequences. This might involve the development of novel recycling and waste management systems for spacecraft and other equipment, as well as the appropriate handling of possible pollutants and other hazardous materials.

Overall, the development of sustainable practices in space travel will be vital for maintaining the long-term survival and profitability of the sector. With the increased interest in space exploration and the development of new technologies like Starship, there is enormous opportunity to stimulate innovation and cooperation in this field and create new ways to sustainable space travel.

The Societal Implications of Starship: What It Means for Humanity and Our Future in Space

The creation and deployment of Starship has huge social ramifications, especially in terms of what it implies for mankind and our future in space.

One major consequence is the potential for Starship to offer up new prospects for space exploration and colonization. With its capacity for long-duration missions and the ability to transport large amounts of equipment and resources, Starship could enable the development of new space habitats and settlements, which could be used to support scientific research, resource extraction, or even the expansion of human civilization beyond Earth.

Another key social aspect of Starship is the possibility for new commercial and economic possibilities in space. With the lower prices and better accessibility provided by Starship, it is probable that we will see more interest in space-based enterprises and initiatives, which may encompass anything from space tourism to asteroid mining.

However, the creation and use of Starship also bring serious ethical and social problems, notably around issues of fairness and access.

It is crucial to ensure that the advantages of space exploration and development are shared equally throughout society, and that the possible dangers and negative repercussions of these activities are thoroughly examined and handled.

Ultimately, the construction of Starship and our future in space pose critical concerns about our position as a species in the cosmos and our obligation to ensuring that we utilize our technical breakthroughs in a manner that benefits all of mankind and the planet we call home.

The Role of Starship in the Exploration and Colonization of Mars

Starship is projected to play a crucial role in the exploration and colonization of Mars, having the potential to change our knowledge of the Red Planet and pave the path for the construction of a sustainable human presence there.

One important feature of Starship for Mars exploration is its enormous capacity and extended endurance. The spacecraft's huge cargo capacity and long-range capabilities will allow it to transfer significant quantities of equipment, supplies, and resources to the surface of Mars, making it feasible to establish a sustained human presence on the planet.

In addition, the totally reusable aspect of Starship makes it a tempting candidate for trips to Mars, since it may considerably cut the cost of launching and maintaining spacecraft. This might make it more possible to send many trips to Mars, allowing for more thorough exploration and study.

Starship might potentially allow the development of new technologies and tactics for Mars exploration and settlement.

For example, the spacecraft may be used to transfer technology and supplies for the generation of in-situ resources on Mars, such as water or oxygen. This might be a significant step in the establishment of a sustainable human presence on the planet.

However, the development of Starship and the colonization of Mars also bring serious ethical and societal problems, notably around issues of fairness and access.

It is crucial to ensure that the advantages of Mars exploration and development are shared equally throughout society, and that the possible dangers and negative repercussions of these activities are thoroughly assessed and handled.

Overall, the creation of Starship is intended to be a huge step forward in the exploration and colonization of Mars, bringing up new potential for scientific investigation, resource extraction, and the establishment of a sustained human presence on the Red Planet.

The Role of Starship in Space Tourism and Commercial Spaceflight

Starship is planned to play an important role in the development of space tourism and commercial spaceflight, enabling new options for people and corporations to reach space and experience the thrill of spaceflight.

One of the primary attractions of Starship for space tourism is its enormous passenger capacity, with the spaceship planned to be able to transport up to 100 people at a time. This might considerably cut the cost of space travel per person, making it more affordable to a larger spectrum of individuals.

In addition, the totally reusable aspect of Starship makes it a tempting candidate for commercial spaceflight, since it may considerably cut the cost of launching and operating spacecraft. This might make it more possible for firms to execute missions in space, such as the deployment of satellites or the development of new space-based technology.

Starship might potentially allow the development of new business possibilities in space, such as the establishment of new space-based industries or the construction of space-based infrastructure. For example, the spacecraft might be used to deliver equipment and materials for the building of new space habitats or the creation of space-based industrial facilities.

However, the development of Starship and the commercialization of space also present serious ethical and social problems, notably around themes of safety, regulation, and equality. It is crucial to guarantee that commercial space operations are handled in a safe and responsible way, and that the benefits of these activities be shared equally throughout society.

Overall, the construction of Starship is projected to be a huge step forward in the development of space tourism and commercial spaceflight, bringing up new options for people and corporations to reach space and stimulate innovation in the sector.

The Future of Spaceflight: What's Next for SpaceX and Starship

The future of spaceflight looks bright for SpaceX and Starship, with the prospect for new and interesting missions in space and the continued development of the spaceship and its capabilities.

One significant area of interest for SpaceX and Starship is the creation of a sustained human presence on Mars. The corporation has previously established plans for a crewed expedition to Mars in the mid-2020s, with the ultimate objective of creating a permanent population on the planet. Starship's high cargo capacity and long-range capabilities make it a tempting alternative for carrying the equipment, supplies, and resources required to accomplish this aim.

In addition, SpaceX is researching additional possible applications for Starship, including point-to-point transport on Earth and the establishment of new space-based businesses and infrastructure. For example, the firm has suggested utilizing Starship to carry people and freight between various spots on Earth in a matter of minutes, transforming the way we travel and conduct business.

Another possible use of Starship is the establishment of a space-based economy, where materials and goods are mined and created in space and brought back to Earth. This might entail the creation of space-based industrial facilities, the mining of asteroids or other celestial bodies, or the development of space tourism and other space-based enterprises.

However, the development of Starship and the future of spaceflight also present fundamental problems concerning regulation, safety, and ethics. It will be crucial to ensure that these challenges are thoroughly studied and handled as we continue to push the frontiers of space exploration and development.

Overall, the future of spaceflight seems bright with the continuous development of Starship and the opportunity for new and fascinating missions in space. As SpaceX and other firms continue to innovate and push the bounds of what is possible, the prospects for space exploration and development are boundless.